WILDLY CUTE ANIMALS!

SEA OTTERS

BY KRISTEN RAJCZAK NELSON

Gareth Stevens PUBLISHING

Please visit our website, www.garethstevens.com. For a free color catalog of all our high-quality books, call toll free 1-800-542-2595 or fax 1-877-542-2596.

Cataloging-in-Publication Data
Names: Rajczak Nelson, Kristen.
Title: Sea otters / Kristen Rajczak Nelson.
Description: Buffalo, NY : Gareth Stevens Publishing, 2026. | Series: Wildly cute animals! | Includes glossary and index
Identifiers: ISBN 9781482471298 (pbk.) | ISBN 9781482471304 (library bound) | ISBN 9781482471311 (ebook)
Subjects: LCSH: Sea otter–Juvenile literature.
Classification: LCC QL737.C25 R35 2026 | DDC 599.769′5–dc23

Published in 2026 by
Gareth Stevens Publishing
2544 Clinton Street
Buffalo, NY 14224

Designer: Jeffrey Taylor
Editor: Kristen Nelson

Photo credits: Series art (tan background) Elena_Design/Shutterstock.com; Cover, p. 1 KensCanning/Shutterstock.com; p. 5 Laura Hedien/Shutterstock.com; p. 7 Erin Underwood/Shutterstock.com; p. 9 elakazal/Shutterstock.com; p. 11 Vladimir Wrangel/Shutterstock.com; p. 13 Kara Capaldo/iStock.com; p. 15 MansonFotos/Shutterstock.com; p. 17 Sean Lema/Shutterstock.com; p. 19 Alaskan Wildlife/Shutterstock.com; p. 21 MichalKrata/Shutterstock.com.

Printed in the United States of America

CPSIA compliance information: Batch #CS26GS: For further information contact Gareth Stevens at 1-800-542-2595.

CONTENTS

Boldface words appear in the glossary.

Ocean Animals

What's that floating on top of the water? Look at it flip over and dive down! Sea otters are marine **mammals**. "Marine" means they live in the ocean. These animals are wildly cute—and wildly important!

Where Are They?

There are three kinds of sea otter. Southern sea otters live off the coast of parts of California. Northern sea otters live in waters near Alaska and Washington State. Russian sea otters live near the coasts of Japan and Russia.

SOUTHERN SEA OTTER

The Sea Otter Look

Sea otters have large heads and small eyes. They have whiskers. Sea otters are covered in fur. Their fur comes in shades of brown. When they get older, some parts of the fur may turn white.

WHISKERS

At Home in the Water

Sea otters have **webbed** feet for swimming. They can close their nose and ears to stop water from flowing in. A sea otter's ocean home may be quite cold. So, sea otters have two **layers** of **dense** fur. This keeps them warm.

Sea otters live along coasts in forests of giant kelp, which is a kind of seaweed. They spend most of their time floating on top of the water. They wrap the kelp around their bodies, so they don't float away!

Togetherness

Sea otters get together and float in huge groups! A group of sea otters is called a raft. Within a raft, male otters commonly stick together. Female otters group up too. Scientists have seen rafts with more than 1,000 sea otters!

Sea Otter Pups

Sea otters have their babies in the water. They are the only kind of otter that does this. Sea otter mothers have one baby at a time. Babies are called pups. When mother otters go looking for food, they wrap pups in kelp on top of the water. This keeps pups in place!

Feeding Time

Sea otters hunt for crabs, snails, sea urchins, and clams on the ocean floor. Some sea otters eat fish too. They float on their backs to eat. If the food has a shell that needs to be opened, a sea otter may use a rock.

Keystone Species

Sea otters are a keystone species. They play a big part in keeping their **ecosystem** balanced. However, these animals are endangered. That means there aren't many left in the wild. And without them, that balance could be lost, causing harm to other living things.

GLOSSARY

dense: Packed very closely together.

ecosystem: A natural community of life that includes living and nonliving features.

layer: One thickness of something lying over or under another.

mammal: A warm-blooded animal that has a backbone and hair, breathes air, and feeds milk to its young.

webbed: Connected by skin.

FOR MORE INFORMATION

BOOKS

Lock, Deborah. *Sea Otters*. New York, NY: DK Publishing, 2023.

Veitch, Catherine. *Meet the National Animals: Fun Animal Facts from Around the World*. London, UK: Happy Yak, 2022.

WEBSITES

Mammals
https://kids.nationalgeographic.com/animals/mammals
Read about more mammals here.

Sea Otters
https://www.montereybayaquarium.org/animals/animals-a-to-z/sea-otter
Find even more information about sea otters here.

Publisher's note to educators and parents: Our editors have carefully reviewed these websites to ensure that they are suitable for students. Many websites change frequently, however, and we cannot guarantee that a site's future contents will continue to meet our high standards of quality and educational value. Be advised that students should be closely supervised whenever they access the internet.

INDEX